GRAB ELECTRONICS

(A Fantastic Electronics Project Manual)

Siddhartha Sinha

pencil

ISBN 978-93-5559-225-5

Published in India 2022 by Pencil

Contributors:
Co-Author: TAPASH SINHA

A brand of
One Point Six Technologies Pvt. Ltd.
123, Building J2, Shram Seva Premises,
Wadala Truck Terminal, Wadala (E)
Mumbai 400037, Maharashtra, INDIA
E connect@thepencilapp.com
W www.thepencilapp.com

Author biography

Siddhartha Sinha was born on 25th November, 1994 in Guwahati (City of Temple) where the Assam's capital Dispur is situated. He was born to Mr.Tapash Sinha, Physics project instructor (Co-author) and Mrs.Uma Sinha, a house-wife.

He completed his schooling as well as Higher Secondary from Modern English School and graduated from the prestigious Arya Vidyapeeth College. The subject which he studied at his graduation was Physics, Geology and Mathematics. His interest in Electronics developed from a very tender age when he observed his father constructing Electronics projects for the school students. He sat alongwith his father in order to learn to construct electronics circuits and wondered how the circuits functioned. As he grew, his interest towards electronics mingled with the idea of philosophy of human life.

Therefore, if you read this book 'Grab Electronics', you would find that in almost all the chapters, he tried to relate the electronics with day to day life in order to make the readers understand that humans are not different than electrons. The character of humans and his life is same as that of electrons.

Even if you dont get the philosophy behind his writings, atleast his way of teaching electronics and circuits in this book will help you clarify your concept.

So, why waste time in wandering, lets begin the show with the idea of Electronics.

CONTENTS

Preface

This book is written with an ardent desire to cherish the world of Electronics and understand the activities of electrons which is the basis for life in Earth. Electrons are present in every atom of this universe. Somewhere I feel that these electrons are may be our ancestors who have been traveling miles and miles and finally formed our earth and then formed us. May be that is the reason I can see lots of similarities between us and electrons. When we build a circuit, the electrons behaves in the same manner as that of humans in the society. When we create a circuit, we are trying to make a world for the electrons where it acts according to our own wish. We can get our desired output for our own benefit just by constructing the right path for electrons. Who knows may be the unseen almighty is also doing the same with us. May be it is creating path for us so that it could achieve some of its desired goal and we are merely wanderers in our life like the electrons wanders in a circuit !!! The way we make a circuit and break it when we dont get the desired output, may be when the unseen power who is running the universe make a path for us and then break and recreate another path just in order to achive its goal !!

It might be anything as it is his story and we are merely actors acting with the help of script installed within us. We are formed and we will be deformed; we are existing now and one day we will be inexistant for definitely other people. But we as an energy will always sustain in this nature, may be with or without our memories and emotions.

I will be talking about this idea later, but for now I think I am addressing to the young minds of the country using my book. And I believe that after reading this book, the students will develop a different perception of electronics and life. They will be able to understand science with the help of the detailed analysis of their day to day life.

Dear readers, my idea is to instill the concept of electronics in a very simple form in your brain. Kindly read each and every sentences wrote in the book; rush to the electronics markets to get the electronic components and other materials and start constructing your world for the electrons who are sitting idle in the conductors.

Chapter 1 Introduction to electronics

Nature is the vast subject to study. I feel it is our responsibility to know the nature as we are actually its part. I'm not the only person who thinks about it; actually the idea was already there with our ancestors. And so, that is the reason why we study different subjects like physics, geology, chemistry, biology, mathematics, etc,. Our ancestors formed these subjects so as to study the great nature through different angles. In the similar way, we studied about electrons and its behavior. Generally, electron is a chemical concept, i.e., it deals with the armature of atom, which is found to be present in the outermost part of an atom. Various scientists like **THOMSON**, **RUDERFORD**, **BOHR**and many more worked to know the secret of electron embedded so technically within the atom. But a new idea approached after a while and that was the movement of electrons through the conductors. That branch which had dealt with movement of electrons is said to be ELECTRONICS.

ELECTORNICS, in simple words, is a branch which studies the behavior of electrons, either in static or in moving form, within the conductors or semi-conductors. It do not deal with the character of electron, i.e., does not describe it. But under "**ELECTRONICS**" we simply study its behavior.

In this chapter we are going to discuss and introduce some terms which is most important to study electronics.

ELECTRIC CHARGE

Electric charge is the fundamental property of particles due to which it attracts another particle of different kind and repel some other particle which are of same kind.

Thus, there are two types of charges which we term as:-

OPPOSITE CHARGE(or **UNLIKE CHARGE**) : A charge, say positive, when attracts another charge of other kind, i.e., negative, are known to be unlike charges.

SIMILAR CHARGES(or **LIKE CHARGES**): A charge, say positive, when repel another charge of same kind, i.e., positive, are known to be like charges.

Thus we say, "**UNLIKE CHARGES ATTRACT EACH OTHER**" and "**LIKE CHARGES REPEL EACH OTHER**".

CONDUCTORS AND INSULATORS

We can do any work when we get permission for doing that. Before entering the classroom in the presence of teacher, we ask, "**MAY I COME IN?**" Why do we ask so? It is actually done so as to take the permission from the teacher and yes, it is regarded as good manner too. It actually depends upon the teacher whether he/she allows us to enter the classroom or not. If he/ she utters the word '**NO**', we cannot enter the room. It is a simple example of granting permission.

Particles do interact with one another which is invisible to us. Granting and asking permission also takes place within the material. Now, we all know that electrons are very much important as it is the only one which can flow within any substance. Electron is like the student who ask permission from the material so as to flow. Those material which grants permission for the flow of electrons is said to be **CONDUCTORS**and those substance which don't permit the electron to flow within itself is said to be **INSULATORS**.

I hope the above definition of conductor and insulator is understood by all students. But, you may not be appreciated for writing the above definition in the exam. So, for the so-called exam point of view, one must remember the definitions as follows:

CONDUCTORS: Those substances through which electricity can flow are called conductors.

E.g.: silver, copper, iron, etc.

INSULATORS: Those substances through which electricity cannot flow are called insulators.

E.g.: glass, paper, rubber, etc.

WHAT IS THE POTENTIAL OF CHARGE?

The above written topic may seem funny as well as interesting, right students? Yes. It is the fundamental nature of every organism to know it's 'capacity' as well as we basically try to judge other's '**capacity**' too. So, how can we ignore 'charge'? Let us understand what I meant by capacity of charge.

Actually by saying '**capacity of charge**', I mean by '**electric potential**'. We define the term as follows:

"***ELECTRIC POTENTIAL AT ANY POINT IN AN ELECTRIC FIELD MAY BE DEFINED AS THE AMOUNT OF WORK DONE IN MOVING A UNIT POSITIVE CHARGE FROM INFINITY TO THAT POINT***".

Let me explain you about the above definition.

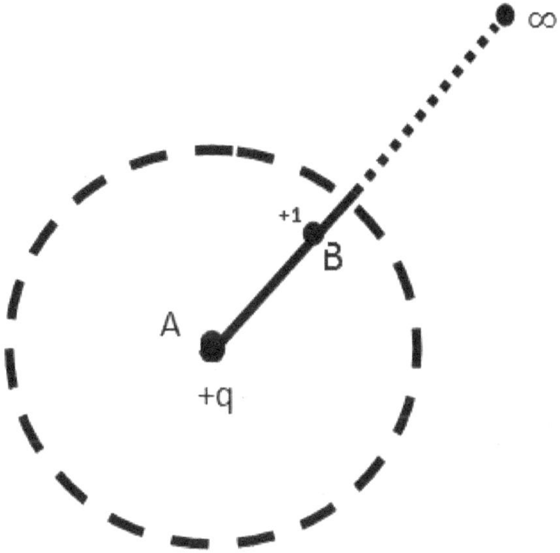

Let us imagine a charge ‘**A**’ (=**+q**) having electric field ‘**E**’ around it.

Let a ‘**+1**’ charge is at infinity (∞). Now, what we do, is, we brought **+1**charge from ∞to the point ‘**B**’. Now, my friends, as we all know that **+ve**is repelled by **+ve**, therefore, if we leave the **+1**charge in the field of another **+ve**charge, it would again move to infinity. Thus, when we bring the **+1**charge from ∞, we do some work, especially when we enter the electric field. Thus, the more strongly **+1**charge is repelled, the more is the capacity of **+q**charge, isn’t it?

It is same as potential in you. If you are more capable, you can defeat all the problems of your path. In the reverse way, we can say that ***the more problems you can solve, the more capable you become***. Similarly, the more a charge is repelled or attracted at any point inside the field of another charge similar or dissimilar, respectively, to it, the more is the potential of the charge having the field.

I hope that your concept is clear.

THE BOUNDARY OF CHARGES

Everything, in this world, have some **limit**or a **boundary**, and accordingly they have got some influencing character.

E.g. - If we consider a Cricket field or a stadium, the players and audiences are influenced by the match. But those who are outside the stadium are not influenced (if we consider there is no radio and television broadcast). They are actually not interested to know what is happening inside the stadium. Thus, the boundary within which the people are influenced by the match is the field of that match.

Similarly, there is also a boundary of charges. Within that boundary if another charge is placed, then it is influenced. Let us consider a +ve charge having a field. Now, if we consider another +ve charge inside the field, obviously it will be repelled, but if –ve charge is placed, it will be attracted.

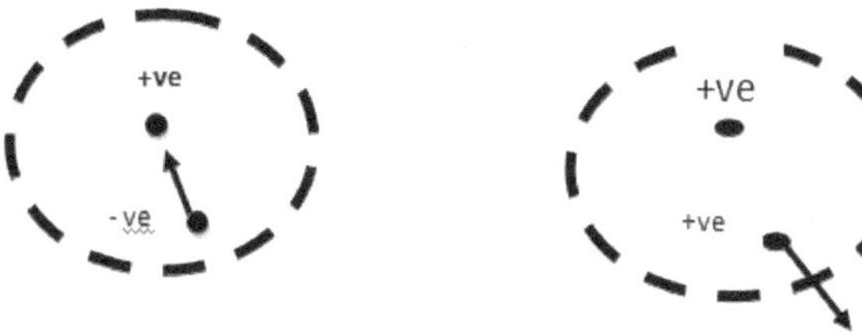

Thus, ***electric field around a charge is the boundary within which if another charge is placed, it experiences a force (either attractive or repulsive).***

POTENTIAL DIFFERENCE

Difference of anything means comparison between two elements, points or anything. We compare generally between the **highest**and **lowest**value. And accordingly we find out the difference between the highest and lowest value. In the similar way, we can also compare the highest and the lowest value of the electric potential. Let us see the below example.

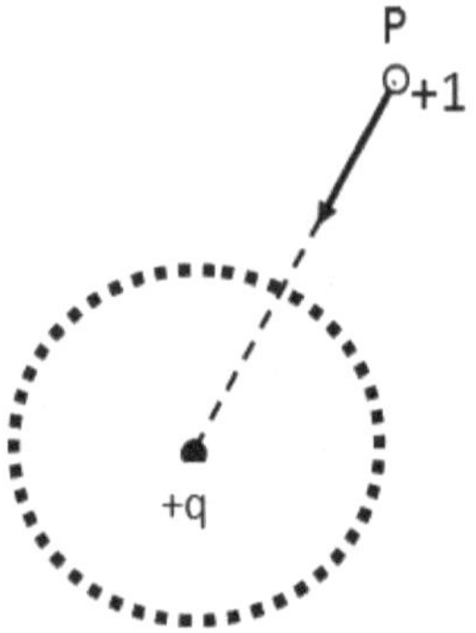

FIG. (A)

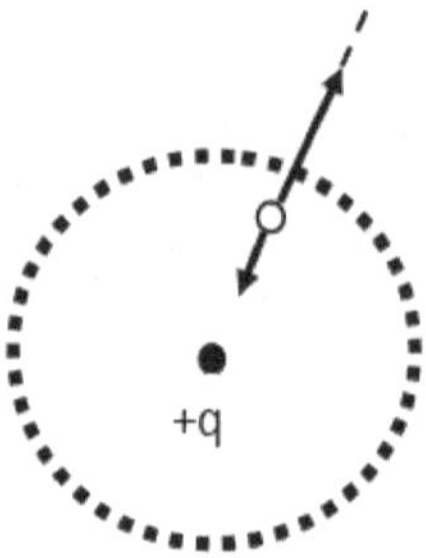

FIG. (B)

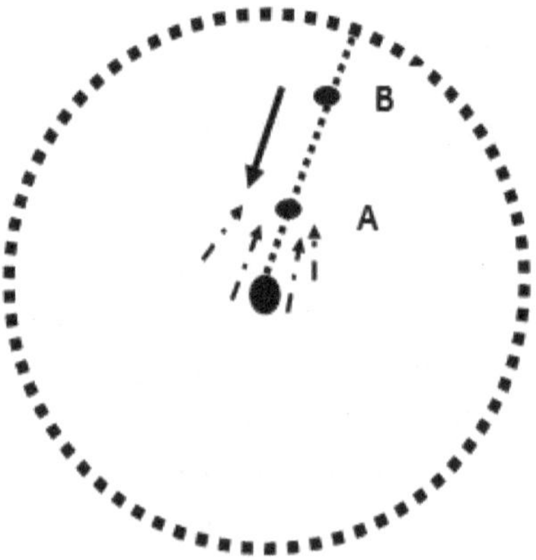

FIG. (C)

Let us consider a {**+q**} charge which has an electric field and at point '**P**' a {**+1**} charge is placed (say at infinity) as shown in **Fig. (A)**. Now we applied force on {**+1**} charge and placed it inside the field of {**+q**} at point '**B**'[**Fig. (B)**]. It experiences some **repulsive force**and we called the work done to be potential of {**+q**} charge. Now, as we take the {**+1**} charge more closer to the {**+q**} charge, it will experience much repulsive force, say at point '**A**' [**Fig. (C)**]. Here, point '**B**' is at lower potential and point '**A**' is at higher potential. Now, if we release the {**+1**} charge, it automatically flows from '**A**' to '**B**'. The same way our rivers flow from **upstream**to **downstream**.

Thus, ***Potential difference across two points may be defined as the amount of work done in bringing a unit positive charge from lower potential point to higher potential point.***

CURRENT

Currentactually depends upon the **potential difference**. When we develop potential difference between the two ends on the conductor, it behaves as if we created a vaccum section in it, due to which the electrons of it starts flowing from negative potential to positive potential. This flow of charge is said to be current.

Thus, ***Current through a conductor may be defined as the amount of flow of charge per unit area.***

Note: When we draw any electric circuit diagram, we mark the flow of current from positive terminal to negative. Such flow of current is said as "Conventional current", as this direction is regarded by convention.

OHM's LAW

According to Ohm's Law, ***the potential difference applied across the two terminals of a conductor is directly proportional to the amount of current flowing through it.***

Mathematically, it is expressed as :

$$V \propto I$$

$$\Rightarrow \quad V = R \times I \quad \ldots\ldots\ldots\ldots\ldots\ldots\ldots \text{(i)}$$

It is observed that when **same amount of voltage**is applied across the two terminals of **conductor made of different elements**but **same length**, the **current flowing**through them are **not same**. This is due to the **obstacle provided to the flow of current**because of the different chemistry of such elements. This property of obstacle is termed as '**Resistance**' and is denoted as '**R**' as represented in the **equation (i)**.

Generally, resistance is present in every conductor. It is even present in insulators too. We might also say that no current flows through insulators because insulators provide much more resistance to the flow of current. But when we make circuits, we use certain special kind of resistors, namely 'Carbon Resistance', which we are going to discuss in the next chapter.

Chapter 2 CARBON RESISTANCE

The character of '**opposition**' is present in all objects of this universe. If you take the example of human civilization, you can clearly understand the fact of opposition. It is actually your and my nature to oppose others when we find that others oppose us, without even knowing whether we're correct or not. The same nature is followed by the electrons. During the flow of electrons, some kind of opposition is offered to its flow. This opposition is termed as Resistance and is denoted by the symbol 'Ω'(OHM).

Now, in order to construct electronic circuits, we use certain man-made resistances known as Carbon Resistance. These resistances are actually used to control the voltage of the circuit by regulating current through it. It can be bought from any electronics market. The carbon resistors look like the **fig. (A)**.

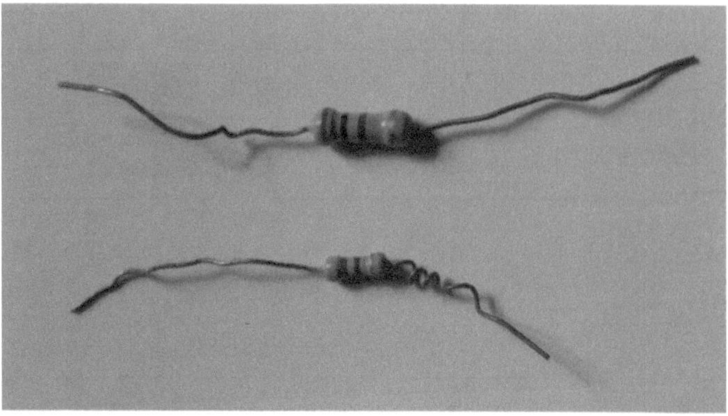

It can be observed from **Fig. (A)**that carbon resistances are decorated with certain colours. These are not merely decorations; rather the colours printed on them designate the power of the resistances. This coding of resistors is important because in such compact components, printed numbers would be a visual challenge for us. Moreover, various circuits require various voltage controlling power. These codes are known as "Resistance Colour Codes". Learning these codes is very essential for an electronics learner and by means of time one must expertise himself in calculating the value of resistance just by seeing the colour printed on it.

There are 10 resistance colour codes as shown in the below table:

NAME OF COLOUR	NUMBER	COLOUR
BLACK	**0**	
BROWN	**1**	
RED	**2**	
ORANGE	**3**	
YELLOW	**4**	
GREEN	**5**	
BLUE	**6**	
VIOLET	**7**	
GREY	**8**	
WHITE	**9**	

TABLE 1: TABLE FOR RESISTANCE COLOUR CODE

These Resistance Colour Codes must be memorized properly so that calculating the value of resistance becomes easier.

We also know that nothing is perfect in this earth. So, how come the man-made resistor be regarded perfect? And that is the case why we require tolerance mark for approximation. The tolerance is also printed in terms of colours and is read as give in the below table:

NAME OF COLOUR	TOLERANCE	COLOUR
NO COLOUR	**±20%**	
SILVER	**±10%**	
GOLDEN	**±5%**	
RED	**±2%**	
BROWN	**± 1%**	

{Note regarding Tolerance mark: Supposing we have a resistance of 10Ω with ±5% tolerance, i.e., Golden Tolerance. It actually means that the resistance that we have is either (10 + 5) = 15 Ω resistor or (10 – 5) = 5 Ω resistor with us. Thus with every carbon resistance, we can find certain tolerance mark which approximates the value of the resistance used in our circuit. }

Now let us observe the calculation of the value of resistance by using the knowledge of Resistance Colour Code.

Before calculating the values of resistance, we must know the relation between the different units of the Resistance, i.e. , Ohms, Kilo Ohms(K) and Mega Ohms(M).

Mathematically, **1 K Ω = 1000 Ω, or, 10^3 Ω**

And, **1 M Ω = 1000 KΩ, or, 10^3K Ω**

i.e., **1M Ω = 1,000,000 Ω, or, 10^6 Ω**

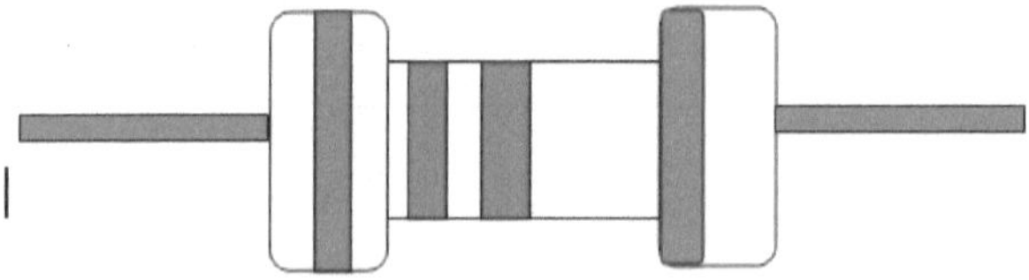

FIG. (B) THE RESISTANCE HAVING THREE SRIPS OF COLOURS- 'RED', 'RED', 'RED' AND A TOLERANCE OF 'GOLDEN'

The carbon resistance is available in the electronics market with four colour strips and also three colour strips. Student must always remember that they should hold the resistance in such a way that the **three or four strips**which are printed **near**to each other must lie towards the **left hand**

sideand the **separated tolerance mark**must lie towards the **right hand side**, exactly as shown in **Fig. (B)**.

Now let us consider the example shown in **Fig. (B)**. The example shown is actually of a three colour stripped resistances, and the strips are coloured as: **Red**, **Red**, **Red**and also consists a Golden tolerance mark. The first strip being '**Red**' is read as '**2**'. **[Note: One must learn the Resistance Colour Code properly in order to find the value of carbon resistances]**. The second strip also being '**Red**' is read as '**2**'. But although the third strip is '**Red**', we don't read the value of the strip as '**2**' rather we read it as the **number of zeroes**as that of the **value of the Code**, i.e., '**00**'. **[If the 3rd strip would have been 'Orange', then instead of '3' we might have written '000']**.

Thus, for the example as shown in **Fig. (B)**, the value of the resistance is **2200Ω**.

I hope the concept is clear now. Even if there remains some confusion, the upcoming execises will clarify your doubt if you go through the chapter properly.

If we need to express the value of the resistance in terms of **Kilo Ohms**, then we must follow the following calculations:

We know that, **1000 Ω = 1K Ω**

∴ by unitary method, we get

$$1\ \Omega = \frac{1}{1000}\ k\Omega,$$

$$\therefore 2200\ \Omega = \frac{1}{1000} \times 2200\ K\Omega$$

$$= \frac{1}{10\not{0}\not{0}} \times 22\not{0}\not{0}\ K\Omega$$

$$= \frac{22}{10}\ K\Omega$$

$$= 2.2\ K\Omega$$

```
      2.2
    ------
10 ) 22
     20
    ----
      20
      20
    ----
      x
```

Hence the overall value of the Resistance shown in **Fig (B)**is **2200 ± 5% Ω**or **2.2 ± 5%K Ω.**

2.2 K Ω may also be written as 2K2. This representation is essential because sometime we might get confused with the place of decimals in the value and also to reduce the chance of getting mistake. Instead of decimal we write 'K' to represent Kilo Ohms. Similarly we can write 'E' in place of decimal representing 'Ohms', also we may write 'M' replacing the decimal from 'Mega Ohm' value.

Q1) Find the overall value of the given resistance.

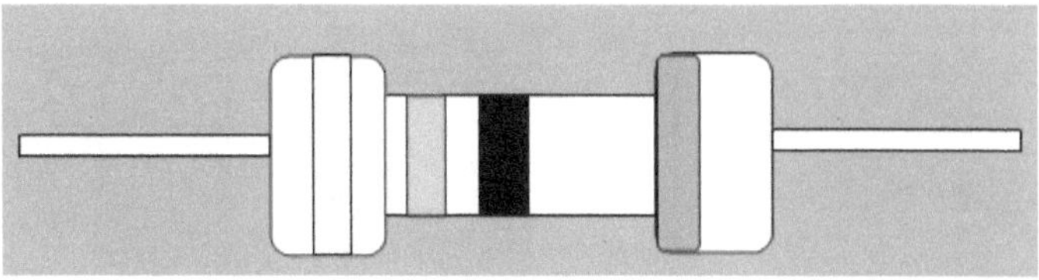

Solution:

This resistance has three colour strips- **Yellow**, **Green**and **Black**; with a **silver tolerance mark**.

Thus, for first strip we will read as '**4**' , for second strip we read as '**5**' and since the third strip is '**Black**' where colour code for '**Black**' is '**0**'. Therefore **we won't read any number for black**. ***[Note: If there is 'Brown' in the third strip, we read it as only '0', for 'Red' it is considered '00', for 'Orange' it is considered '000' and so on.]***

Basically, the value of this resistance is **45 Ω**. But with tolerance silver we need to apply **±10%**

Thus, the value of this resistance if 45 ±10% Ω,i.e., either the resistor is having a value of 55 Ω or 35 Ω

Q2) Find the overall value of the given resistance.

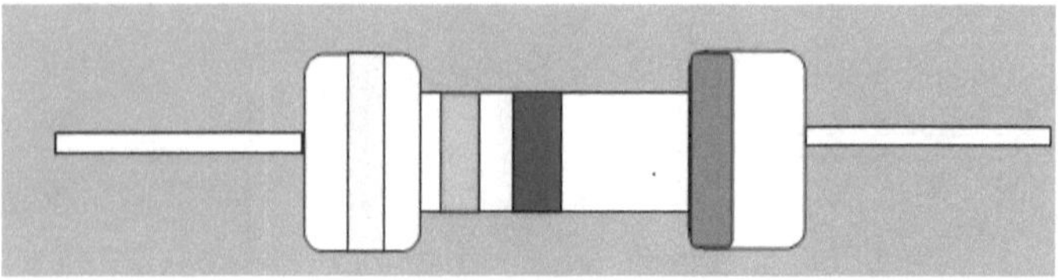

Solution:

This resistance is same as the resistance as shown in '**Q1**', except the third strip which is coloured '**Brown**'. Thus the first strip is read as '**4**', second strip is read as '**5**' and third strip is read as '**0**'. Therefore, the value of this resistance is **450 Ω**.

With tolerance mark, the value of this resistance is read as 450 ± 5% Ω.

Tolerance mark is generally not considered for practical purpose. When we make any circuit, we generally tell the value of the resistance as 450 Ωonly rather than 450 ± 5% Ω The tolerance is nothing but simply denotes the accuracy. So, when we make any circuit, we don't give preference to the tolerance mark rather only colour codes. Also, one must not get

confused while understanding the colour codes of resistance and the tolerance mark.

Q3) Find the overall value of the given resistance.

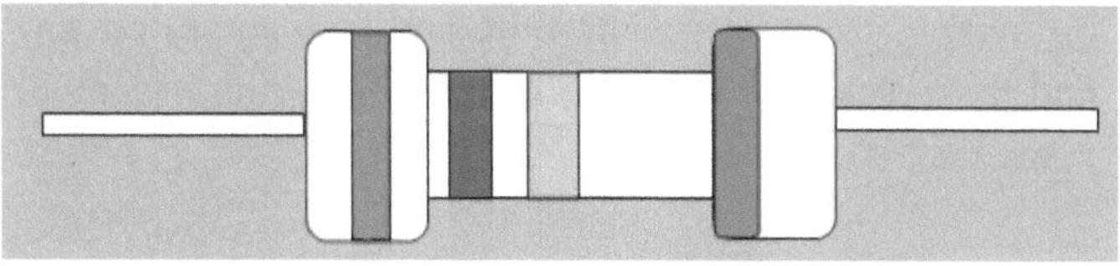

Solution:

The Colour Code of the resistance as represented in the diagram is as follows:

RED -> 2

VIOLET -> 7

GREEN -> As the code of this colour is '5' and since it is the third strip, so it is read as '00000'.

Therefore, the value of the given resistance is 2,700,000 Ω, or, 2.7M Ω.

I wish that all of you have understood how to find the value of a carbon resistance by observing the colour printed on it. And remember my dear students that if you master yourself in finding the value of the resistance, half of your work is done because you can construct any circuit according to your need by applying appropriate resistor in it.

So, now visit any electronic shop and grab some resistors and try to calculate the value of it by observing it's colour code. Isn't it interesting?

There are many more components which we need to know about viz. **capacitors**, **transistors**, **IC's**, **transformer**, **diodes**etc. so that we can construct different circuits of different functionality as per our requirement. Unlike the Carbon resistance, we are going to understand about these components while learning about the circuits.

Before starting to construct the circuits, let us know how shall we prepare our project report (**specially for the students of class XII**) in the next chapter.

Chapter 3 Report writing for Physics project for Class XII

I don't believe the fact that one should try with some easy subject or try to construct a simple circuit first and then try to construct the complicated one. Learning never takes place in this manner. In order to understand the concept, first interest towards the subject is important. **If you have true interest, then you will never bother about the level of complicacy.**Trying to construct the complicated circuit becomes same as constructing or trying the simple ones if you focus on acquiring knowledge. But yes **if you think to make excuses more than learning, then this book is not for you**.

Although this book and the chapters are made for those who are interested to learn about electronics, this chapter is dedicated to Class XII students who struggle to make the report presentable to examiner at the time of practical exam. It is the basic format which one needs to follow after the completion of construction of the circuit.

The format is as follows:

1. Cover of the report (as shown in the next page)
2. Content
3. Certificate
4. Acknowledgement
5. Aim of Study
6. Introduction
7. Circuit diagram
8. Materials required
9. Working Principle
10. Conclusion

From Chapter 4 onwards, as we will discuss about the circuits, Introduction, circuit diagram, materials required, working principle and conclusion will be written in details so that along with making the circuit, you will be able to understand the concept behind its working.

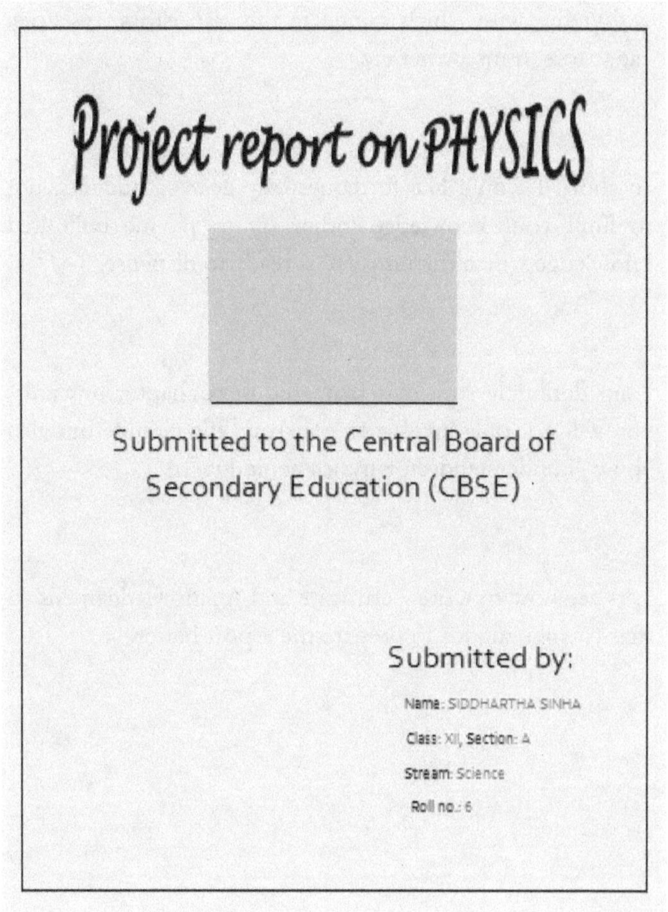

Project report on PHYSICS

Submitted to the Central Board of Secondary Education (CBSE)

Submitted by:

Name: SIDDHARTHA SINHA

Class: XII, Section: A

Stream: Science

Roll no.: 6

Now, what are you waiting for? Let us pursue the knowledge of different circuits which you can try to construct. Before starting the further chapters, I would like to ask you not only to imitate the circuit as per circuit diagram but also after observing its working status, please try to understand the internal function of the electrons within the wire which connects the transistors, resistors, capacitors, transformer etc.

In short, I would like to request my dearest students, not to limit your knowledge rather try to pursue unlimited knowledge which the universe is ready to dispense.

I am definitely sure that from the next chapter onwards, you will not only be able to construct the circuits but also you will understand the physics behind it.

Lets see how to write Certificate and Acknowledgement so that you get an idea to prepare the report better.

Acknowledgement

As a student of class XII, I was required to do project of my choice and as advised by our physics teacher (Teacher's name), I have chosen the project "Name of the project".

I acknowledge with regards and extend my sincere thanks to our physics teacher (Teacher's name), without whose help and full co-operation, the completion of this project would not have been possible. His ability and timely guidance has been a source of inspiration to me.

(Name of the student)

(Name of the school, Address)

Certificate

This is to certify that (Student's name) roll no. .... of class XII has satisfactorily completed the project "Name of the project" during the academic year 20__ - __ under my guidance has been satisfactory.

DATE:

(Name of the teacher)

PHYSICS DEPARTMENT

(School name, Address)

Signature of Supervisor	Signature of Principal	Signature of External

This is the way your teacher certifies you for the extraordinary effort which you give the whole year in order to prepare your project and you acknowledge your beloved teacher without whose support reaching any milestone is not possible.

So, since we have got the idea of how to prepare our project report, why to wait? Lets begin our lessons.

Chapter 4 Electric bell

INTRODUCTION :

Electronics has always been a very interesting subject to be studied. The characteristic of every electronic component is different and it becomes a joy to study these characters. Resistance has the character to resist the flow of current and regulate the voltage depending upon the necessity of the circuit. Similarly transistor acts as a gate for the electric flow and with its three terminals it help to maintain the current in such a way that it provides an audio amplifier stage to the circuit. Excessive current flows through the collector of the transistor whereas very minimum current flows through the base and the current through the base and collector is emitted out through the emitter of the transistor. So, in this manner every component has different functions. The electronic circuit acts just as any physical community where everybody has different work to accomplish. Hence electronics is a very interesting subject to be studied.

Therefore I've thought to construct a very simple but interesting circuit of 'Electric Bell' where I've tried to show the beauty of electronics. I've used several components like transistor, resistor, L.E.D, Buzzer, wire and a push switch, and organized them in such a way that they establish a very beautiful community of electrons and completes the circuit.

The social behavior within the circuit does not happen with intelligence rather it actually happens because of the character of different components used there.

Let us now understand the function of the circuit.

CIRCUIT DIAGRAM :

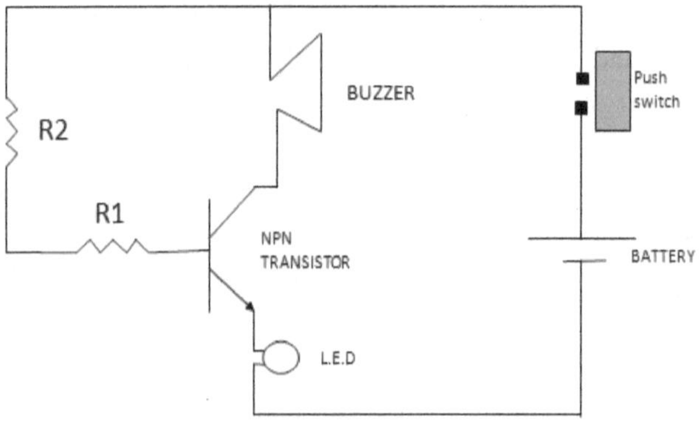

Materials Required:

1. NPN Transistor (BC548)
2. Resistors R1(1.2k Ω) and R2(12k Ω)
3. L.E.D
4. Buzzer
5. Push switch
6. Battery (9V)
7. Copper wire
8. Bread board

WORKING PRINCIPLE :

The working of the circuit is very simple.

I've used a NPN transistor in this circuit. Transistor is a semiconductor device which is used to control the flow of electric current within the circuit with the help of the BASE of the transistor. In order to know how, let us proceed with the working principle.

The connection is shown in the circuit diagram. When we press the Push switch which is connected in the path between the battery and the transistor, current starts flowing throughout the circuit. Maximum current flows through the collector of the transistor and through the base very minimum current flows. The current from the collector and the base are emitted out through the emitter of the NPN transistor. Since the buzzer is connected to the collector and a L.E.D is connected to the emitter, the electron flows through them, thus producing a buzzing sound and glowing the L.E.D. The resistors that are connected regulates the voltage across the NPN transistor, otherwise excessive current may damage the transistor permanently. Thus, resistance acts as a protector of the transistor in this circuit.

In this way the circuit works.

Conclusion :

This is an amplifier circuit. We have seen the various characteristics of different components that work as a community of electronic world. Moreover utilizing the concept of amplification, we may construct many such audio amplifier circuits. We may also construct Automatic lights, Burglar alarm, and many other alarms. This is only possible due to the NPN transistor.

Chapter 5 Automatic Street Light

Introduction:

Every organism in this earth receives and consume several forms of energy. Solar Energy is one the common form. It is our technology which has developed **Transducer**so as to **convert one form of energy into another**. We have converted the mechanical energy into electrical energy which is proved to be very useful for human lives. We can definitely convert the form of energy but inefficiently. Energy is already lost during this process. But we often over utilize it for which too much energy loss takes place. We can observe in our daily life that even during daytime the street light remain glowing. In this manner a lot of energy is lost. In order reduce this loss, we may use an '**Automated Technology**' in those street lights if possible. Thinking about this fact I've thought about a theme named "**AUTOMATIC STREET LIGHT**".

The circuit is similar to "**Electric Bell**" but here we can see transistor functioning as a switch.

In order to know how, let's understand how the circuit works.

Circuit diagram:

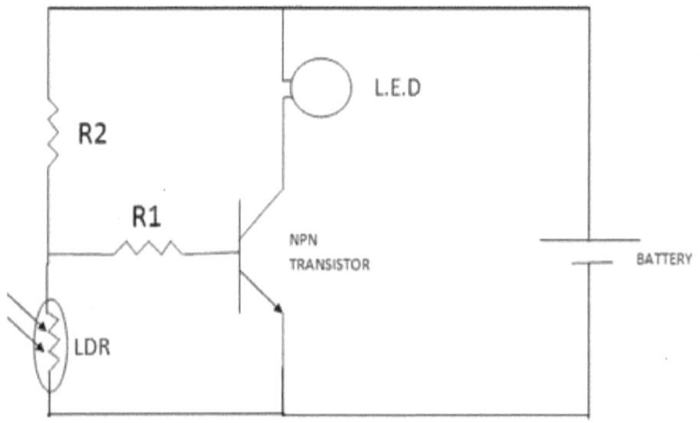

Figure 1

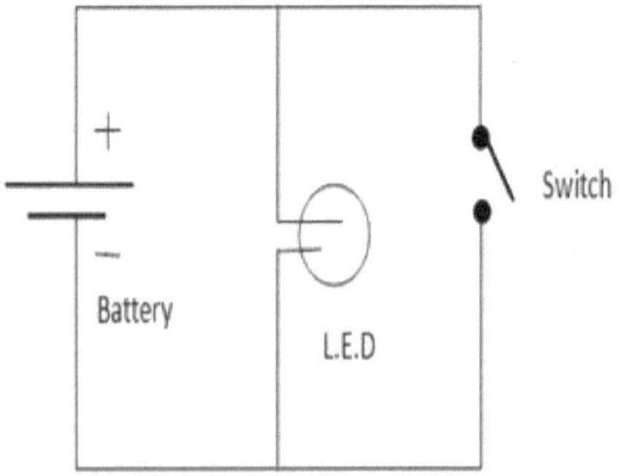

Figure 2

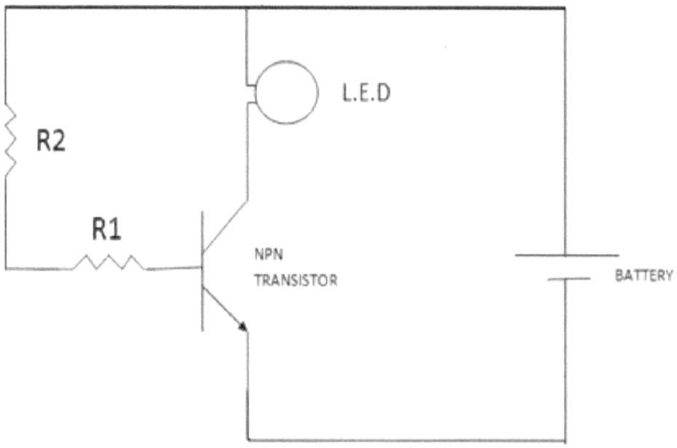

Figure 3

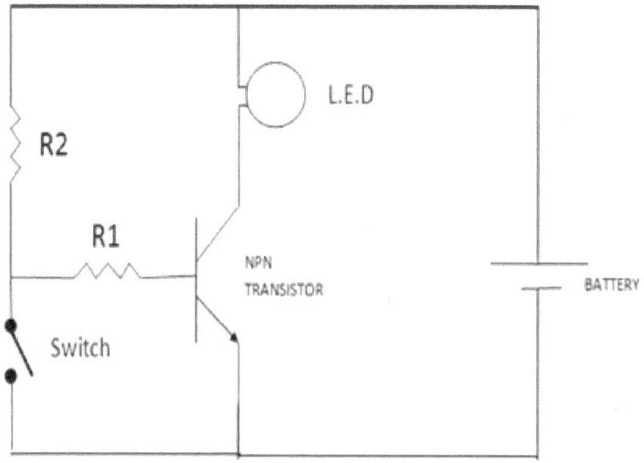

Figure 4

Materials Required:

1. NPN Transistor (BC548)

2. Resistors R1(1.2k Ω) and R2(12k Ω)

3. L.E.D.

4. L.D.R.

5. Battery (9 V)

6. Copper wire

7. Bread board

WORKING PRINCIPLE:

Figure 1in the circuit diagram is the actual circuit of '**Automatic Street Light**' which is constructed.

In order to understand the actual working of **Figure 1**, we need to first study the concept behind **Figure 2**, **Figure 3**and **Figure 4**.

The circuit shown in **Figure 2**is very simple. A **battery**, **LED**and a **switch**is used to make this circuit. Unlike any ordinary circuit, as we observe, the switch is connected **parallel**to the **battery**and **LED**. Actually, when switch is **OFF**, electrons from the battery passes through the **LED**due to which the **LED**glows. But when we **ON**the switch, **LED**stops glowing. It is because, the switch is a conductor whereas **LED**is a **semi-conductor**. As electron find its easiest way through the conductor (switch) rather than semi conductor (LED), as a result of which the LED don't glow. This phenomenon is termed as **short-circuit**.

Figure 3is the main portion of **Figure 1**as observed. What exactly happens here is that electrons from the negative terminal of the battery flows through the **emitter**of the transistor and maximum electron proceeds through the **collector**, whereas very less amount of electrons flows out through its **base**. Due to the same reason we consider **base**to be the **output**of the transistor as the **current signal is amplified at the base**. Since the circuit is completed and the **LED**is in the path of collector, it glows.

Now if we combine the **Figure 2**and **Figure 3**, we get the resultant circuit as **Figure 4**. In this circuit, we can observe that between the **base**and **emitter**, a switch is connected. As discussed before that switch is made with a conductor which gives the **easiest way for the electron to flow**, when we **ON**the switch, maximum electrons flow through it rather than the transistor. Very less amount of current flows through the transistor which is not sufficient to glow the **LED**. Hence when we **ON**the switch, **LED**don't glow and if we **OFF**it, the circuit functions as in **Figure 3**and thus the **LED**glows.

Our main circuit is same as **Figure 4**but instead of switch we have used an **LDR**(**Light Dependent Resistance**) which functions when sunlight or light from any source falls on it. Let me Summarize the working of the actual circuit.

In this circuit a **N-P-N TRANSISTOR**is used. **NPN TRANSISTOR**consist of three tags – **Base(b)**, **Collector(c)**and **Emitter(e)**. When the source (battery) is connected to the circuit, current flows throughout it. Electron flows through the **Emitter**and **maximum electron**moves out through the **Collector**, whereas very **less electrons**pass out through the **Base**. At the **base**the **current is amplified**. Since the **LED**is present in the passage of the **COLLECTOR**and **EMITTER**through which the electron is flowing, the LED glows.

Now carefully if the circuit is observed, the **LDR**(**Light Dependent Resistance**) is connected in between the **Base**and **Emitter**. We must remember that **LDR**works under the action of light, i.e., ***ELECTRON GETS ITS PASSAGE WHEN LIGHT FALLS ON LDR***. The **resistance of LDR decreases upto approximately 100Ω when light falls in it**, whereas **when no light fall on it,**

the resistance of it increases upto approximately 10MΩ. Hence, when **no light**falls on LDR, it does not work, i.e. **it behaves as an insulator**. Now when we supply the light energy to the LDR, the transistor section is **short-circuited**. As there is no **LED**in the passage of **LDR**, no light glows. Whereas when we stop the supply of light to the **LDR**, the electrons again moves through the **LED**via the **transistor**, thus glowing the **LED**.

In this manner, the circuit works.

CONCLUSION:

This circuit can be used in many ways. Since this circuit can carry and support the voltage supply of 3v up to 24v. This circuit may be used as 'Automatic night lamp', 'Laser communication circuit' as a security alarm circuit, 'Transistor as a switch' etc. utilizing it's semi-conduction character. Solar energy makes this circuit quite interesting. Also by making few changes to the connection we may alter it's functioning and we may use it as solar generator too. Moreover this circuit actually behaves as one of the logic gates – "**NOT Gate**".

This is a very sensitive and interesting circuit. Isn't it?

So, if your teacher or examiner asks you to make a circuit of "**Transistor as a Switch**", then you may right away go with the circuit shown in **Figure 4**without any hesitation.

Let us now study about the logic gates in the next chapter.

Chapter 6 The Logic Gates

Introduction:

Logic, as per dictionary, is defined as **'the art or science of reasoning'**. Thus, Logic Gate may be defined as the **entry**or **exit**on the basis of the reasoning. It is exactly the **'VISA PASS'**, if the US Embassy disapproves our VISA, we can't travel to USA and if they approve it, we can definitely travel. It is merely a statement of '**yes**' and '**no**'. Most prominently, if we speak (in terms of electronics), **according to the input we set, we must get a desired output**. There are generally, three standard logic gates, viz., **OR Gate**, **AND Gate**and **NOT Gate**. In our syllabus, certain combinations are also discussed. The combination of **OR**and **NOT**gate produces **NOR**gate and combination of **AND**and **NOT**gate produces **NAND**gate.

Generally, logic gates are primarily implemented using diodes and transistors that acts like electronic switches.

Diode: A diode is a two terminal electronic component consisting of **P-type**and **N-type**semiconductor plate (where '**P**' plate and '**N**' plate are in junction) which basically allow the current to pass only through one direction. Thus, diode can also act like a switch. Whenever, current would travel in one desired direction, diode would conduct. Whereas, if current tries to flow in opposite direction, it acts like Switch **OFF**condition and will not conduct electricity through it.

Transistor: Transistor is a three terminals semiconductor device consisting either two **N-type**plates and one **P-type**plate; or two **P-type**plates and one **N-type**plate, where the three terminals are termed as **Collector**, **Base**and **Emitter**. If the **base**is **P-plate**and other two terminals are **N-plates**, then the transistor is **NPN transistor**. While, if the **base**is **N-plate**and other two terminals are **P-plate**, then the transistor is **PNP transistor**. **NPN transistor**acts as an **amplifier**as well as **switch**. The **Base**and **Emitter**are always in **forward bias**; whereas **Base**and **Collector**are in **reverse bias**. Thus, if somehow we can **short circuit**the **Base**and **Emitter**, then the transistor can act as a **switch**.

The actual working of the circuit can be well understood when we will construct circuit and see how it functions.

Circuit diagram, Working and Truth Tables of different logic gate circuits:

In order to understand the working principle of logic gates, we need to observe the "Truth Table" for each gates.

A) OR Gate

A	B	Y
0	0	0
1	0	1
0	1	1
1	1	1

TABLE 1: TRUTH TABLE OF 'OR' GATE

The truth table of OR gate is shown in Table1. The Zero (0) and One (1) represented in the truth table means switch is 'OFF' and 'ON' respectively. From the Table1 it must be clear that in OR Gate, in order to get any output we need to Switch ON either switch 'A' or 'B'. If all the input remains OFF, then we won't get the output too. But the absence of any one input will not affect the output i.e., if switch 'A' is ON while keeping the switch 'B' OFF, we will still receive the output 'Y'.

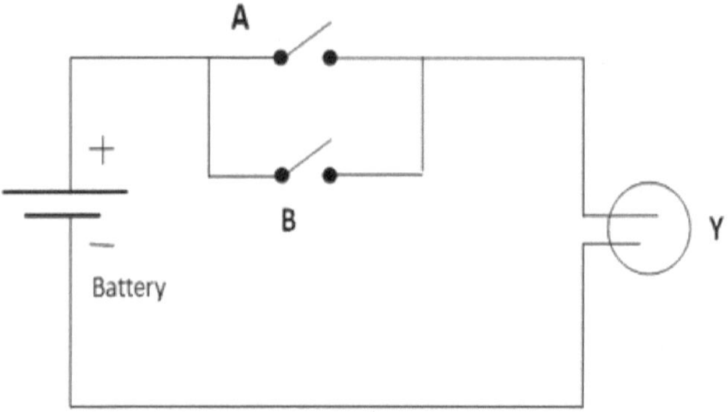

Figure1 : Circuit of OR Gate

Let us understand it through a simple circuit.

As we can see in the Figure 1 that switches 'A' and 'B' (Connected in Parallel) are OFF, i.e., they are in zero (0), hence there is no current flow in the circuit as a result of which the output 'Y' is also zero (0) {which means that the bulb does not glow}. Let us now switch ON the switch 'A'. What is your observation?

Definitely, when the Switch 'A' is switched ON, current flow from the battery throughout the circuit, irrespective of Switch 'B' being OFF. Due to the flow of electrons, the bulb will also glow giving us the output as ON ,i.e., One (1). Similarly, let us switch OFF the Switch 'A' and ON the Switch 'B'. The current will again flow through the circuit via 'B' this time thus glowing the bulb resulting the output One (1).

I hope you are able to understand the Logic of 'OR' gate from this circuit. Now in this circuit if we connect diode D1 and diode D2 in forward biased condition, the function of the OR Gate circuit will remain the same.

Your question might be why diode? Right? The answer is simple. It is because your teacher or examiner may not like this very simple circuit of OR Gate rather wants you to construct a circuit using diode.

The diagram of **OR gate**using diode is shown below.

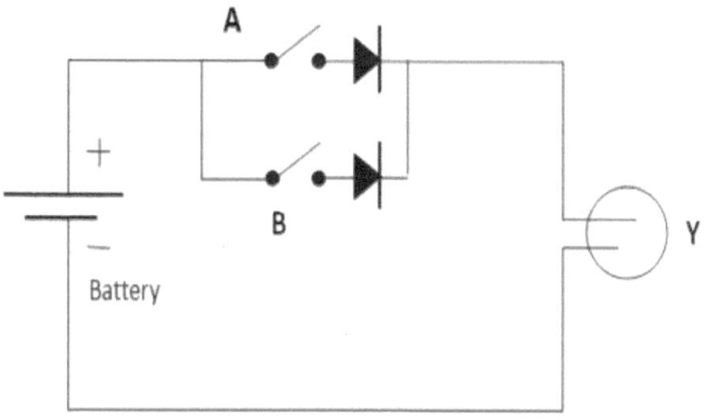

Figure2 : Circuit of OR Gate using Diode

In this way by simply connecting rectifier diodes with Switches '**A**' and '**B**', you can prepare your project of **OR Gate**using diode as shown in **Figure 2**.

B) AND Gate

A	B	Y
0	0	0
1	0	0
0	1	0
1	1	1

Table 2 : Truth Table of AND Gate

The truth table of AND gate is shown in **Table2**. From the **Table2**it must be clear that in **AND**Gate, in order to get any output we need to Switch **ON**both switch 'A' and 'B'. If all the input remains **OFF**, we will definitely not get the output too. The absence of any one input will affect the output i.e., if switch 'A' is ON while keeping the switch 'B' OFF, we will not receive the output 'Y'.

Let us understand it through a simple circuit.

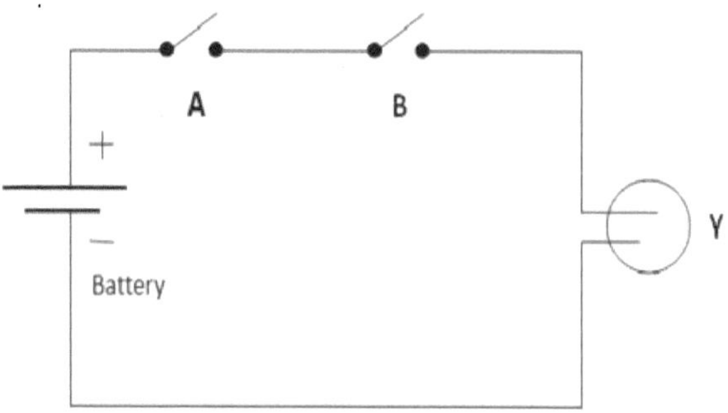

Figure3 : Circuit of AND Gate

As we can see in the Figure 3 that switches 'A' and 'B' (Connected in Series) are OFF, i.e., they are in zero (0), hence there is no current flow in the circuit as a result of which the output 'Y' is also zero (0) {which means that the bulb does not glow}. Let us now switch ON the switch 'A'. What is your observation? When the Switch 'A' is switched ON, current do not flow throughout the circuit as Switch 'B' is also switched OFF. As a result, we do not receive any output, i.e., Output 'Y' is zero (0). Similarly, let us switch OFF the Switch 'A' and ON the Switch 'B'. The current will not flow through the circuit via 'B' this time as Switch 'A' is already OFF, thus resulting the output as Zero(0) again. Now, we will receive the output 'Y' as One (1) only when we will ON both the switches 'A' as well as 'B'.

I hope you are able to understand the Logic of 'AND' gate from this circuit. Now in this circuit if we connect diode D1 and diode D2 in forward biased condition, the function of the AND Gate circuit will remain the same.

The diagram of AND gate using diode is shown below in Figure 4.

In this way by simply connecting rectifier diodes with Switches 'A' and 'B', you can prepare your project of AND Gate using diode as shown in Figure 4.

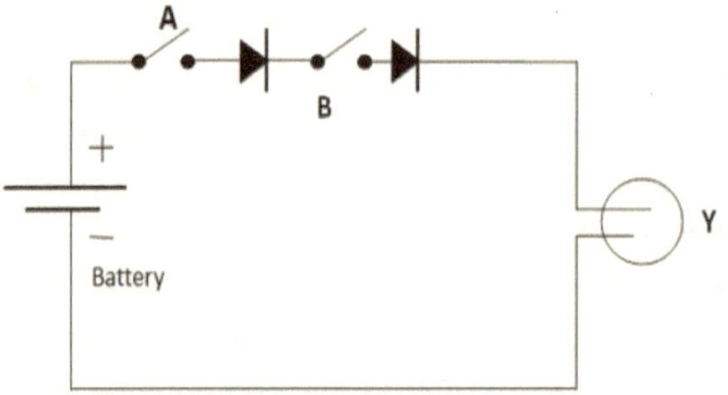

Figure 4 : Circuit of AND Gate using diode

C) NOT Gate

Y	Y'
0	1
1	0

Table 3: Truth Table of NOT Gate

From **Table 3**we can observe that whenever we provide any input, output remains absent, whereas when we do not provide any input, we get an output. Thus, NOT Gate is also regarded as '**INVERTER**'.

Isn't it interesting? Let us learn about it by constructing a very simple circuit using battery, switch and LED as shown in Figure 5.

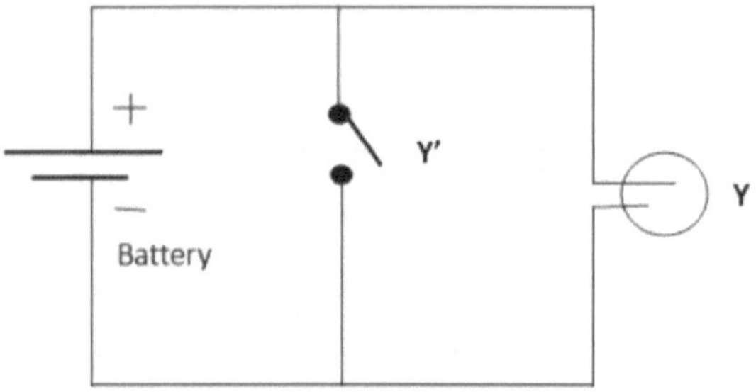

Figure 5 : Circuit of NOT Gate

Before understanding the working of this circuit, I would like to ask you the observation. I am asking so because if you recall you had already seen this circuit in **Chapter 5**.

You are right! Good Job!This circuit is the best example of **short circuit**as by using the switch we can **shorten**the **path**of the **flow of electrons**. We had observed that when the **Switch Y**' is **ON**, electrons gets an easier path to travel through it rather than the LED. Hence, when the input **Y'**is **One (1)**, the output **Y**is **Zero (0)**. Similarly, when we switch **OFF**the Switch **Y'**, current flows through the LED as this is the only path the electrons can flow. Therefore, **when there is Zero (0) input the output is One (1)**. Isn't it inverting the situation? So, can't we term NOT Gate as an '**Inverter**'?

Now, definitely when you are making a project for Class XII final examination, this simple circuit will not be acceptable. Your teacher or examiner will ask you to make a NOT Gate circuit using **transistor**. And here we start to panic. Because, we feel transistors to be a very difficult concept. But if you have read **Chapter 5**precisely, I am sure that you are enough confident to use transistor to construct any circuit. Try to recall the concept we read in the previous chapter and those who are confused after reading this statement, I would like to ask you to please go through the Chapter 5 thoroughly.

Now, I am presenting the circuit diagram of **NOT**gate using **NPN transistor**and the working principle is requested to be consulted from Chapter 5 of **Figure 4**.

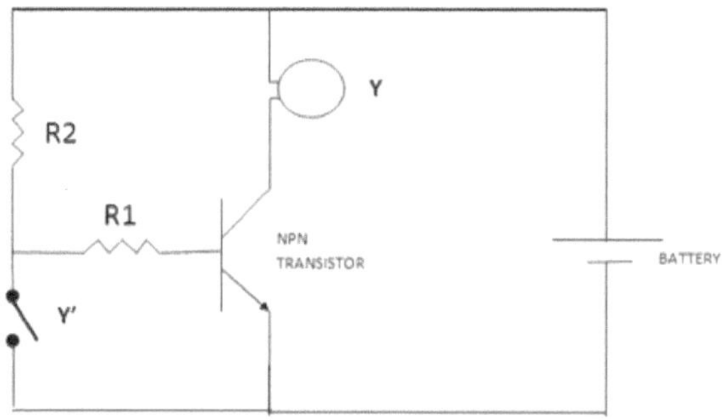

Figure 6 : Circuit of NOT Gate using Transistor

Let us now tally Truth table of NOT Gate with Figure 6. Are you able to get the observation? If not, then check the connection properly.

D) NOR Gate

The Truth Table of **NOR Gate**is nothing but the combination of **OR Gate**and **NOT Gate**. The truth table of **OR gate**and **NOT gate**is shown in **Table1**and **Table3**respectively. The Zero (0) and One (1) represented in the truth table means switch is **'OFF'** and **'ON'** respectively, as discussed earlier. From the discussion it must be clear that in OR Gate, in order to get any output we need to Switch ON either 'A' or 'B'. If all the input remains OFF, then we won't get any output. But the absence of any one input will not affect the output.

From Table3 we can observe that whenever we provide any input, output remains absent, whereas when we do not provide any input, we get an output. So, NOT Gate is also regarded as **'INVERTER'**.

Now, in order to construct **NOR Gate**, we must use **OR Gate as input**and **NOT gate as output**. Let us see the truth table for NOR Gate.

A	B	Y'	Y
0	0	0	1
1	0	1	0
0	1	1	0
1	1	1	0

TABLE 4: TRUTH TABLE OF NOR GATE

The truth table explains the circuit entirely. We need to just follow the truth table, to verify whether our circuit is correct or not. Any step we use, must match the output characteristic of the table, i.e., when we keep any of the switches, i.e., **A**or **B OFF**while the other **ON**(as a result of which the **Y'**is **ON**because it is the **parallel**output of **A**and **B**), then the output **Y**must remain **OFF**as per Truth Table. If the circuit made does not work according to the truth table, then we must consider that there is a fault in the circuit.

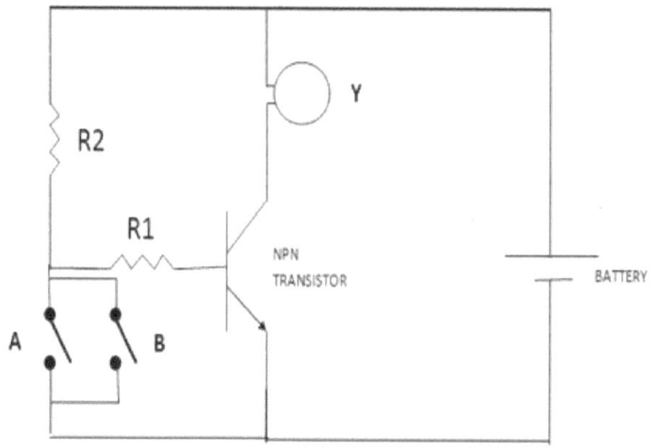

Figure 6 : Circuit of NOR Gate using Transistor

I am not just going to explain the working of Figure 6 once again. Try to study the working from Chapter 4 and use the concept of parallel connection of switches 'A' and 'B' in the place of LDR.

D) NAND Gate

The Truth Table of **NAND**Gate is nothing but the combination of **AND**Gate and **NOT**Gate. The truth table of **AND** gate and **NOT**gate is shown in **Table2**and **Table3**respectively. From the discussion it must be clear that in **AND**Gate, in order to get any output we need to Switch **ON**both '**A**' and '**B**'. If all the input remains **OFF**, then we won't get any output. The absence of any one input will affect the output. From **Table3**we can observe that whenever we provide any input, output remains absent, whereas when we do not provide any input, we get an output. So, **NOT**Gate is also regarded as '**INVERTER**'.

Now, in order to construct NAND Gate, we must use AND Gate as input and NOT gate as output. Let us see the truth table for NAND Gate.

A	B	Y'	Y
0	0	0	1
1	0	0	1
0	1	0	1
1	1	1	0

TABLE 4: TRUTH TABLE OF NAND GATE

The truth table explains the circuit entirely. We need to just follow the truth table, to verify whether our circuit is correct or not. Any step we use, must match the output characteristic of the table, i.e., when we keep any of the switches, i.e., **A**or **B OFF**while the other **ON**(as a result of which the **Y'** is **OFF**because it is the series output of

Aand **B**), then the output **Y**must remain **ON**as per Truth Table. If the circuit made does not work according to the truth table, then as we know, we must consider that there is a fault in the circuit.

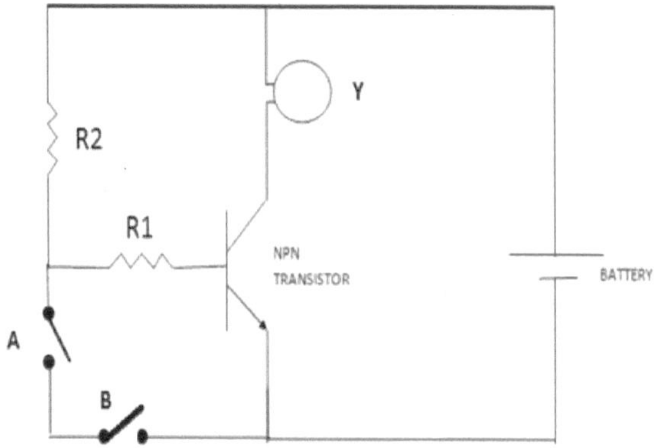

Figure 7 : Circuit of NAND Gate using Transistor

I ask you again regarding the working of **Figure 7**as I did in case of **Figure 6**. ***Tell it aloud or write it somewhere so that you can remember it better.***

CONCLUSION:

As we have seen the wonderful design of the above circuits of logic gates which is actually based on the data of the truth tables. Also, one more point we can understand is that, if we have a good observation power, we can use the same circuit for the different purposes.

Isn't it interesting? You will find it more interesting once you construct these circuits by yourself. Let us proceed ahead with the next chapter where we will learn regarding the 'Audio Amplifier Circuit'.

Chapter 7 AUDIO AMPLIFIER CIRCUIT

INTRODUCTION :

There are many amplifier circuits present in the electronic market. Amplifier is actually used in order to increase the amplitude of sound or sometimes may be of current too. Depending upon the circuits, the components such as NPN transistor and PNP transistor are used may be in a qualitative and quantitative manner. We may also find some kind of Alarm circuits which are also amplifiers. Moreover certain Integrated Circuits (ICs) also act as an amplifier.

Although a single NPN transistor act as an amplifier circuit, but I've tried to construct a fine amplifier circuit using NPN and PNP transistor. Let us understand how it works.

CIRCUIT DIAGRAM :

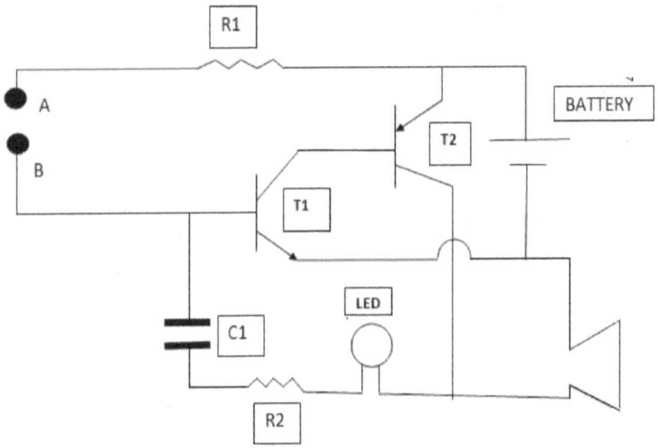

Materials Required :

1. NPN Transistor (BC548) and PNP transistor (BC558)
2. Ceramic Capacitor C1 (0.01 pF) ; pF is pico farad
3. Resistors R1(320 k Ω) and R2(1k Ω)
4. Battery (9V)
5. Speaker

6. Copper wire

7. Bread board

WORKING PRINCIPLE :

When the two terminals of a battery is connected to the circuit, current flows throughout it. Current flows thought the emitter of the PNP transistor (T2) where the current is suddenly amplified at the base. It seems as if the PNP transistor (T2) attains power stage and supplies power to the NPN transistor (T1).T1 provides audio-amplifier stage to the circuit.

The current flows through the intervening space of base of T1 which amplifies the current signal. This happens because at the base of T1 the velocity of the electrons increases. As a result, the sound can be heard through the speaker. This sound is nothing but an indication of Audio amplification. Here, Capacitor and resistors are used in order to filter the sound and to control the voltage, respectively.

But the above working is not possible until and unless we touch the terminals 'A' and 'B'. Since we used 9V battery (comprising 90mA),we don't get sound when we touch the terminals. In order to produce a good quality sound, minimum 500mA to 1A current is required. So, I've connected a colouring LED so as to regulate the voltage which produces sound in a variable manner.

In this way the circuit works.

CONCLUSION :

This is an amplifier circuit. So utilizing the concept of amplification, we may construct many such audio amplifier circuits. We may also construct Automatic lights, Burglar alarm, Water level indicator, Rain alarm, Fire alarm and many other alarms. This is only possible due to the NPN transistor.

This is a very beautiful circuit we discussed in this chapter. In this circuit, every component has their individual role. Capacitor here filters the current supply so that we receive a clear and amplified sound through the speaker. If we replace this capacitor of 0.01 pF with 10 mF(micro farad) electrolytic capacitor, then we get a sound equivalent to our heart beat. Doesn't it sound interesting? It really happens. Try it and you will be amazed by the magic of electronics. Now if you use 20 mF or 30 mF capacitor, the beat could be fastened.

What I would like from my dearest students is that don't just copy the circuit from circuit diagram. Rather experiment with the component like capacitor to understand the concept of it. Even by increasing or decreasing the value of capacitor, you will find differences in the sound produced by the speaker because of the charging and discharging character of the capacitor.

As we knew about how the transistors, resistors and capacitors work in the circuit, let us learn about another interesting circuit named "Burglar Alarm".

Being your teacher, I would request you again to try to understand the practical concept of electronics circuit. The circuit which you will make in the next chapter is same as that of this chapter, but we are simply going to use the concept of short circuit in the next chapter. I hope you are ready for the next trip towards the new invention. Lets leave the unnecessary talks and proceed ahead towards the next chapter.

Chapter 8 BURGLAR ALARM

INTRODUCTION:

Today we have seen many problems related to the security of the self and the property. Of course, we must 1st concentrate in the security of the self but we must also try to protect our property from the cruel and poor Burglars. We earn and without our permission someone takes the whole property is actually intolerable. So it is our Prime duty to think smart as this society is going smarter. There are many circuits in the electronic market that helps to make us secured. But the circuit which I've constructed is very sensitive that if a very few amount of light falls on the sensor then the output may not be provided. So, I've provided an external source to it and it can be fitted in the entrances through which there is the total possibility through which a burglar may enter, so that if any unknown person at unwanted time enters our house then suddenly alarm will be given out. The circuit is very sensitive.

Let us know how the circuit works.

CIRCUIT DIAGRAM:

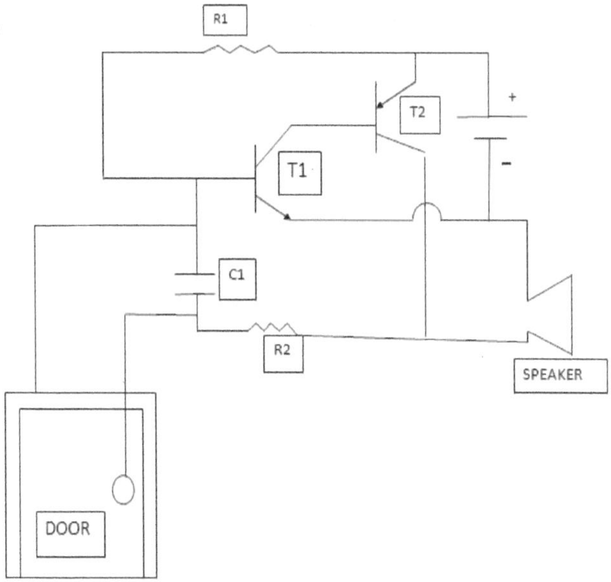

Materials Required :

1. NPN Transistor (BC548) and PNP transistor (BC558)
2. Ceramic Capacitor C1 (0.01 pF) ; pF is pico farad
3. Resistors R1(320 k Ω) and R2(1k Ω)
4. Battery (9V)
5. Speaker

6. Copper wire

7. Bread board

WORKING PRINCIPLE:

When the two terminals of a battery is connected to the circuit, current flows throughout it.PNP transistor attains power stage and supplies power to the NPN transistor .NPN transistor provides audio-amplifier stage to the circuit. In this circuit main role is played by NPN transistor. The current flows through the intervening space of base of NPN transistor which amplifies the current signal. This happens because at the base the velocity of the electrons increases. As a result, the sound can be heard through the speaker. This sound is nothing but an indication of Audio amplification. Here, Capacitor and resistors are used to filter the sound and to control the voltage, respectively.

But, since a wire from the two ends of the ceramic capacitor is connected to the door; therefore the flow of the current is short-circuited. This clearly means that maximum current flows through the wire rather than the capacitor because for electrons, choosing the conductor would be easier than a capacitor. As a result, the sound is not given by the speaker because the required amount of current isn't flowing through it. As soon as someone opens the door, the short circuit connection is broken, the capacitor takes its charge filtering the current flowage and thus the alarming sound is produced as usual.

In this manner the circuit works.

Conclusion:

This is an amplifier circuit. So utilizing the concept of amplification, we may construct many such audio amplifier circuits. We may also construct Automatic lights, Water Level Indicator, and many other alarms. This is only possible due to the NPN transistor.

In my bag of electronics, I have another beautiful presentation ready to be presented in front of you. In the next chapter, you will be learning about the difference of function ability between NPN and PNP transistor.

Chapter 9 DIFFERENCE BETWEEN NPN AND PNP TRANSISTOR

INTRODUCTION:

Semiconductors have really made our life easier. The digital gadgets we talk about are only because of these semiconductors. While studying the chapter of semiconductor I've gone through the transistors. After I studied about NPN Transistor and PNP Transistor, I got a great interest regarding the fact. First I shall talk something about the NPN and PNP Transistor and then I would talk regarding the circuit (that actually consist of two parts) which I have constructed and from this circuit we will be able to make out the difference between the two.

Actually, NPN is one of the two types of bipolar transistors, consisting of a layer of P-doped semiconductor (the "base") between two N-doped layers. A small current entering the base is amplified to produce a large collector and emitter current. That is, when there is a positive potential difference measured from the emitter of an NPN transistor to its base (i.e., when the base is high relative to the emitter), the transistor becomes active. In this "ON" state, current flows between the collector and emitter of the transistor. Most of the current is carried by electrons moving from emitter to collector. To allow for greater current and faster operation, most bipolar transistors used today are NPN because electron mobility is higher than hole mobility(***Please refer NCERT PHYSICS CLASS XII book to get the Electron and hole concept in P-doped and N-doped semi-conductor.***)

Moreover, PNP Transistor consists of a layer of N-doped semiconductor between two layers of P-doped material. A small current leaving the base is amplified in the collector output. That is, a PNP transistor is "**ON**" when its base is pulled low relative to the emitter. In a PNP transistor, emitter-base region is forward biased. So electric field and carriers will be generated. They should flow towards the base junction, but the base part is very thin and has low conductivity. The reverse biased collector base part has generated holes. So due to the electric field, carriers or electrons get pulled by the holes.

The arrows in the NPN and PNP transistor symbols are on the emitter legs and point in the direction of the conventional current flow when the device is in forward active mode.

Now let us understand how the circuit works.

CIRCUIT DIAGRAM:

CIRCUIT 1: Circuit consisting NPN Transistor :

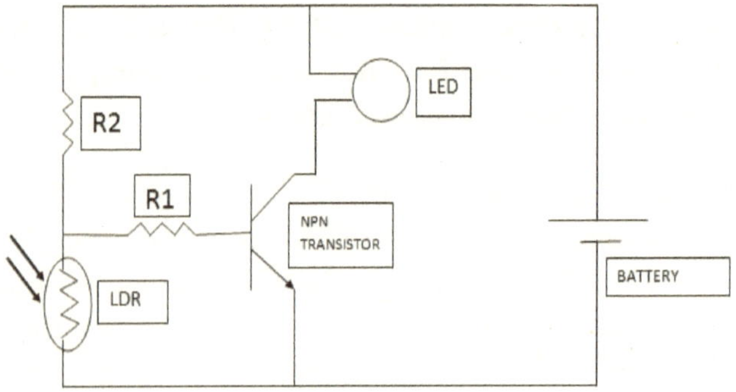

CIRCUIT 2: Circuit consisting PNP Transistor.

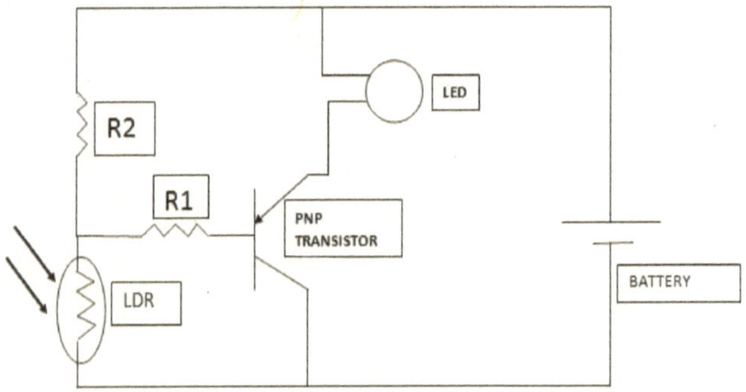

Materials Required :

1. NPN Transistor (BC548) and PNP Transistor (BC558)
2. Resistors R1(1.2k Ω) and R2(12k Ω)
3. L.E.D.
4. L.D.R.
5. Battery (9 V)
6. Copper wire
7. Bread board

WORKING PRINCIPLE:

CIRCUIT 1: Circuit consisting NPN Transistor:

The circuit consisting NPN transistor attains audio amplifier stage. Actually what happen here is that when battery is connected to the circuit, electrons gets a route to flow through the emitter of the NPN Transistor and maximum electron flows out through the collector; and a very small amount of electron moves out through the base. Actually it happens because; the Base-emitter is in forward bias. The voltage gets amplified at the base. Since, the electrons move through the LED, it glows indicating Switch 'ON' condition.

But when the light falls on LDR, current gets an easiest way to flow through it and thus very small amount of electron enters through the emitter. As a result sufficient voltage is not attained by the LED and thus it does not glow. It is in Switch 'OFF' mode.

CIRCUIT 2: Circuit consisting PNP Transistor:

The circuit consisting PNP Transistor attains Power Stage. The collector is in reverse bias condition with the base as we can observe in the circuit. When battery is connected to the circuit, electrons don't get a way to flow, neither through the PNP transistor nor through the LDR. As a result, the LED connected to the emitter doesn't glow.

But when light falls on LDR, electrons gets conducted through it. Since, Base- emitter is in forward bias, some amount of electrons flows though the base to the emitter. But this small amount is amplified at emitter and it seems that PNP transistor provides power to any circuit (through it's base) it is connected to. In this case too, electrons are conducted though the emitter. Since LED is present is the passage of emitter, it glows. So from this explanation, we have understood that when light do not falls on LDR, PNP Transistor is in Switch 'OFF' mode and when light falls on LDR, the LED also glows.

In this way the circuits work.

CONCLUSION:

Very interestingly, now if you take both the circuits to light, 1st circuit will be switched 'OFF' and 2nd circuit is 'ON'. Again, when we take both the circuits to dark, 1st circuit will glow and the second gets switched 'OFF'. In this way both the circuits will show a "FLIP-FLOP" action. In other words, we may also say that this circuit acts as a manual multivibrator. It seems quite interesting.

This is an amplifier circuit. So utilizing the concept of amplification, we may construct many such audio amplifier circuits. We may also construct Automatic lights, Burglar alarm, and many other alarms. This is only possible due to the NPN transistor. Also using PNP transistor we can make a very strong amplifier circuit, utilizing it's "power stage" as we had seen in the previous two chapters.

Chapter 10 Transistor as an amplifier

INTRODUCTION:

We had already seen how transistor works as amplifier in Chapter 7. We are going to make another circuit to demonstrate Transistor as an amplifier.

Transistor is a semi-conductor. As the name suggest, it does not work perfectly as conductor. We may think it to be its disadvantage, but actually speaking this is a great advantage of a transistor that due to this character, we are able to use it as an amplifier. Today whatever amplifier we use or talk about is actually a result of transistor. Without transistor, this would have never been possible. Transistor works by the movement of the charge carriers- may be majority or minority. By doping the semi-conductor in such a material with more or less atomic no., we receive p- or n- plate respectively.

I've used N-P-N Transistor in which Emitter and Collector is N- plate while Base is the P-plate. Where Base is the Output. The working of the circuit I've constructed is written below.

Lets us now understand how does it works.

CIRCUIT DIAGRAM:

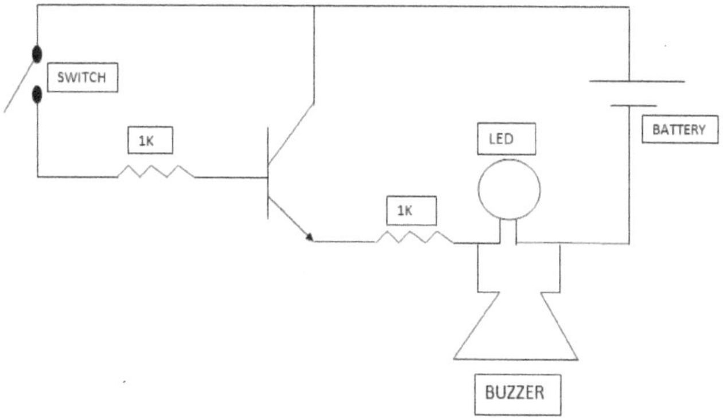

Materials Required :

1. NPN transistor (BC548)
2. Resistors 1k Ω
3. Battery (9V)
4. Buzzer
5. LED
6. Copper wire
7. Switch
8. Bread board

WORKING PRINCIPLE :

The working of the circuit is very simple.

I've used a NPN transistor in this circuit. The connection is shown in the circuit diagram. When we connect the battery to the circuit, electrons start flowing throughout the circuit. Electrons passes through the emitter and maximum electrons flows out through the collector. But very small amount of electrons flows out through the base. Since the base of the transistor is very small, the electron emitted out attains a velocity in comparison to the electron coming out of the battery. As the velocity of the electron increases, it means that the distance of it from the mean position also increases. In other words we may say that the electric signal is amplified. Thus, Transistor act as an amplifier providing Amplifier stage to the circuit. Signal emitting out through it is amplified. Hence we can receive an amplified sound through the buzzer.

In this way the circuit works.

Conclusion :

This is an amplifier circuit. So utilizing the concept of amplification, we may construct many such audio amplifier circuits. We may also construct Automatic lights, Burglar alarm, and many other alarms. This is only possible due to the NPN transistor.

Chapter 11 FIRE ALARM

INTRODUCTION :

It is being mentioned in our Vedas that everything is made with the five forms of matter- Air, Water, Earth, Aether and Fire. And it is not an incorrect tell. All the elements are equally important in all our lives. But anything in access, may lead to a danger. E.g.- air is important for us to survive, but if we face excessive amount of air, then we may die even due to our heart blockage; similarly water which is the most essential part of everyone's life, if received in an excessive amount may lead to death. So the case of fire. Fire if once catches, it becomes too much difficult to normalize it. So if we become careful, then of course we can save our lives as well as property from any such drastic disaster caused because of fire.

Thinking about it, being a science student, I've prepared a circuit, which will alert us from such disaster to happen. Let us now understand the science of an amplifier circuit and observe its usage.

CIRCUIT DIAGRAM :

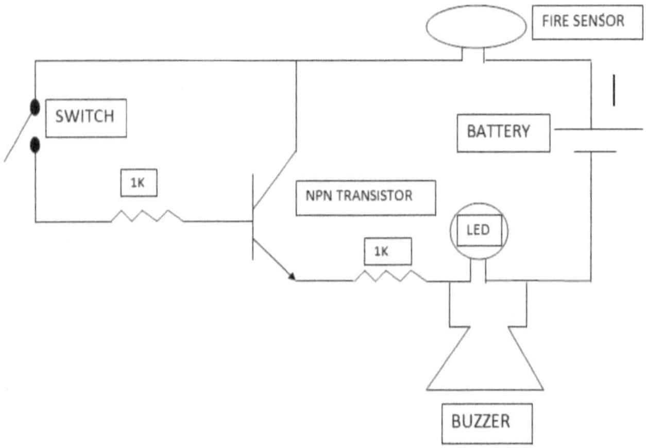

Materials Required :

1. NPN transistor (BC548)
2. Resistors 1k Ω
3. Battery (9V)
4. Buzzer
5. LED
6. Copper wire
7. Switch
8. Bread board

WORKING PRINCIPLE :

The working of the circuit is very simple.

I've used a NPN transistor in this circuit. The connection is shown in the circuit diagram. When we connect the battery to the circuit, we can't observe any kind of action in the circuit. It is because the circuit is incomplete. As soon as the fire touches the fire sensor, electrons start flowing throughout the circuit. Electrons passes through the emitter and maximum electrons flows out through the collector. But very small amount of electrons flows out through the base. Since the base of the transistor is very small, the signal emitting out through it is amplified. Hence we can receive an amplified sound through the buzzer.

In this way the circuit works.

Conclusion :

This is an amplifier circuit. So utilizing the concept of amplification, we may construct many such audio amplifier circuits. We may also construct Automatic lights, Burglar alarm, and many other alarms. This is only possible due to the NPN transistor.

As from the Chapter 4 onwards, we had enough studied as well as seen the working of the Transistor, I feel let us take a break and again we will get back to transistors.

Let us now explore the Chapter 12 which will demonstrate the charging and discharging capacity of Capacitor.

Chapter 12 Current Discharge Capacitor

Introduction :

Capacitor is an electronic component that stores electric charge. The capacitor is made of two close conductors (usually plates) that are separated by a dielectric material. The plates accumulate electric charge when connected to power source. One plate accumulates positive charge and the other plate accumulates negative charge.

The capacitance is the amount of electric charge that is stored in the capacitor at voltage of 1 Volt.

The capacitance is measured in units of **Farad (F)**.

The function of the Capacitor can be explained theoretically but the practical application looks very magical and we can say that Electronics can create magic. The main element of the magic is the Capacitor and how the magic works, we will study this through this simple circuit named 'Current Discharge Capacitor'.

Let us see, how the circuit works.

Circuit Diagram:

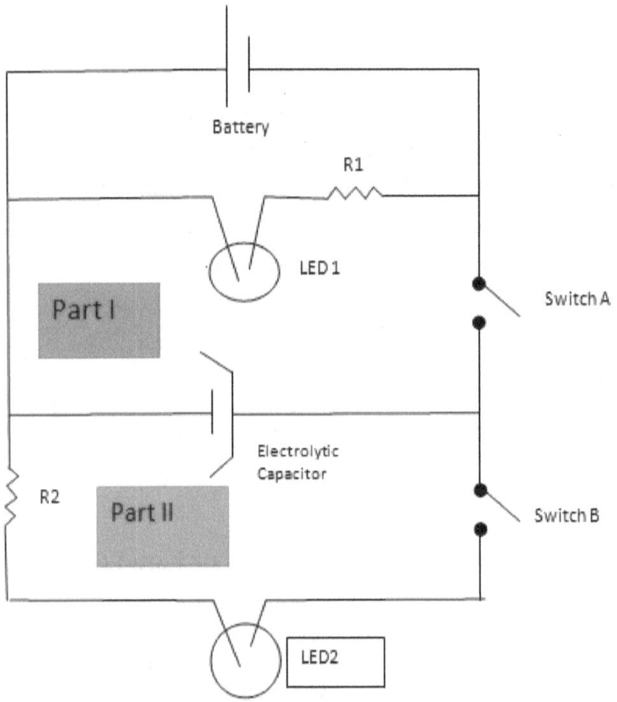

Materials Required :

1. Electrolytic Capacitor (2000 mF or 3000 mF)
2. Switches
3. LED
4. Battery
5. Resistors (1k Ω)
6. Copper wire
7. Bread board

Working Principle:

The working of the circuit is very simple.

The circuit can be divided into two parts – **Part I**and **Part II**. **Part I**consists of **LED**1, resistor **R1**and Switch **A**. While **Part II**comprises **LED2**, resistor **R2**and Switch **B**. Electrolytic capacitor is commonly connected to both the Parts as shown in diagram. Moreover, a Battery, i.e., the power source has been connected to the circuit.

When Switch A is ON, current flows through resistor 1 , electrolytic capacitor and LED1 thus completing the Part1 of the circuit. In this scenario, Capacitor is getting charged as current flows through it. Since the battery is connected parallel to the LED1, it remains glowing.

Switching OFF the Switch 'A', when we ON the switch 'B', Part II of the circuit becomes active and the current is discharged from capacitor, thus glowing the LED2. This happened because the Plates inside the capacitor are charged when current was flowing through it. Whereas, when the power supply went Off, Capacitor acted as battery, discharging current from it in order to glow the LED2.

Resistors are used in the circuit in order to regulate the current within the circuit and also because to prevent the LED's from being damaged.

In this manner the circuit works.

Conclusion:

Capacitors are used to store electrical charge. They are a basic component of electronics. The most common use for capacitors is energy storage. Additional uses include power conditioning, signal coupling or decoupling, electronic noise filtering, and remote sensing. Moreover, Capacitors are also used in the circuit of Fullwave or Bridge rectifier in order to filter out the AC pulse from the rectified DC. Because of its varied applications, capacitors are used in a wide range of industries and have become a vital part of everyday life.

Thanks for reading

www.ingramcontent.com/pod-product-compliance
Lightning Source LLC
LaVergne TN
LVHW100335181026
850499LV00039B/1090
* 9 7 8 9 3 5 5 5 9 2 2 5 5 *